The Botanical Magazine, Vol. 1 Or, Flower-Garden Displayed

William Curtis

Imprint

This book is part of TREDITION CLASSICS

Author: William Curtis
Cover design: Buchgut, Berlin – Germany

Publisher: tredition GmbH, Hamburg - Germany
ISBN: 978-3-8424-8356-9

www.tredition.com
www.tredition.de

Copyright:
The content of this book is sourced from the public domain.

The intention of the TREDITION CLASSICS series is to make world literature in the public domain available in printed format. Literary enthusiasts and organizations, such as Project Gutenberg, worldwide have scanned and digitally edited the original texts. tredition has subsequently formatted and redesigned the content into a modern reading layout. Therefore, we cannot guarantee the exact reproduction of the original format of a particular historic edition. Please also note that no modifications have been made to the spelling, therefore it may differ from the orthography used today.

THE

Botanical Magazine;

OR,

Flower-Garden Displayed:

IN WHICH

The most Ornamental Foreign Plants, cultivated in the Open Ground, the Green-House, and the Stove, are accurately represented in their natural Colours.

TO WHICH ARE ADDED,

Their Names, Class, Order, Generic and Specific Characters, according to the celebrated Linnæus; their Places of Growth, and Times of Flowering:

TOGETHER WITH

THE MOST APPROVED METHODS OF CULTURE.

A WORK

Intended for the Use of such Ladies, Gentlemen, and Gardeners, as wish to become scientifically acquainted with the Plants they cultivate.

By WILLIAM CURTIS,

VOL. I

"A Garden is the purest of human Pleasures."

Verulam.

CONTENTS

[1] — Iris Persica.

[2] — Rudbeckia purpurea.

[3] — Helleborus hyemalis.

[4] — Cyclamen Coum.

[5] — Erythronium Dens Canis.

[6] — Narcissus Minor.

[7] — Cynoglossum Omphalodes.

[8] — Helleborus Niger.

[9] — Iris pumila.

[10] — Anemone Hepatica.

[11] — Erica herbacea.

[12] — Dodecatheon Meadia.

[13] — Coronilla glauca.

[14] — Primula villosa.

[15] — Narcissus Jonquilla.

[16] — Iris variegata.

[17] — Cactus flagelliformis.

[18] — Geranium Reichardi.

[19] — Hemerocallis Flava.

[20] — Geranium Peltatum.

[21] — Iris Versicolor.

[22] — Nigella damascena.

[23] — Tropæolum majus.

[24] — Agrostemma coronaria.

[25] — Dianthus chinensis.

[26] — Stapelia variegata.

[27] — Convolvulus tricolor.

[28] — Passiflora cœrulea.

[29] — Reseda odorata.

[30] — Lilium chalcedonicum.

[31] — Jasminum officinale.

[32] — Mesembryanthemum dolabriforme.

[33] — Aster tenellus.

[34] — Browallia elata.

[35] — Crepis barbata.

[36] — Lilium bulbiferum.

INDEX. — Latin Names

INDEX. — English Names

PREFACE.

The present periodical publication owes its commencement to the repeated solicitations of several Ladies and Gentlemen, Subscribers to the Author's Botanic Garden, who were frequently lamenting the want of a work, which might enable them, not only to acquire a systematic knowledge of the Foreign Plants growing in their gardens, but which might at the same time afford them the best information respecting their culture—in fact, a work, in which Botany and Gardening (so far as relates to the culture of ornamental Plants) or the labours of Linnæus and Miller, might happily be combined.

In compliance with their wishes, he has endeavoured to present them with the united information of both authors, and to illustrate each by a set of new figures, drawn always from the living plant, and coloured as near to nature, as the imperfection of colouring will admit.

He does not mean, however, to confine himself solely to the Plants contained in the highly esteemed works of those luminaries of Botany and Gardening, but shall occasionally introduce new ones, as they may [Pg 004] flower in his own garden, or those of the curious in any part of Great-Britain.

At the commencement of this publication, he had no design of entering on the province of the Florist, by giving figures of double or improved Flowers, which sometimes owe their origin to culture, more frequently to the sportings of nature; but the earnest entreaties of many of his Subscribers, have induced him so far to deviate from his original intention, as to promise them one, at least, of the Flowers most esteemed by Florists.

The encouragement given to this work, great beyond the Author's warmest expectations, demands his most grateful acknowledgements, and will excite him to persevere in his humble endeavours to render Botany a lasting source of rational amusement; and public utility.

Botanic Garden,
Lambeth-Marsh,
1787.

[1]

Iris Persica. Persian Iris.

Class and Order.

Triandria Monogynia.

Generic Character.

Corolla 6-partita: Petalis alternis, reflexis. Stigmata petaliformia.

Specific Character and Synonyms.

> IRIS *Persica* corolla imberbi, petalis interioribus brevissimis patentissimis. *Linn. Syst. Vegetab. p. 79. Sp. Pl. p. 59.*
>
> IRIS bulbosa præcox minus odora Persica variegata. *Moris. hist. 2. p. 357.*
>
> XIPHIUM Persicum. *Miller Dict. ed. 6. 4to.*
>
> The Persian bulbous Flower-de-luce. *Parkins. Parad. p. 172.*

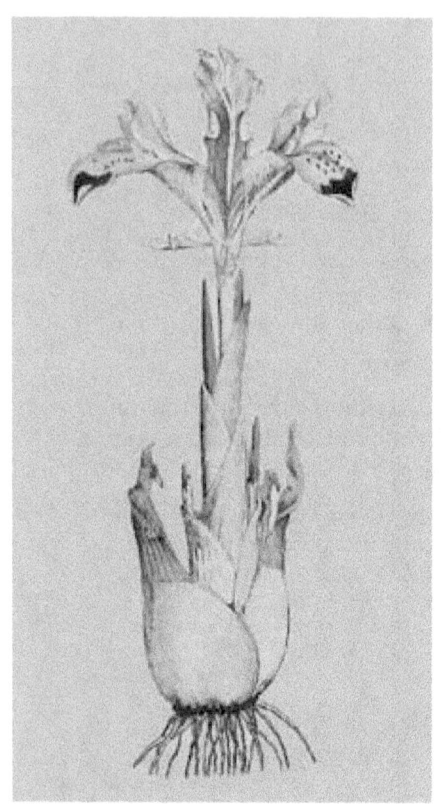

A native of Persia. Flowers in February and March. Its beauty, early appearance, and fragrant blossoms, make it highly esteemed by all lovers of flowers; like the Hyacinth or Narcissus it will blow within doors in a water-glass, but stronger in a small pot of sand, or sandy loam; a few flowers will scent a whole apartment: it will also blossom in the open air, but requires warmth and shelter; it is propagated by offsets and seeds; the best flowering roots are imported from Holland, they bear forcing well; and hence this plant may be had to flower a full month or six weeks in succession.

Parkinson remarks, that in his time (1629) it was very rare, and seldom bore flowers.

[Pg 007]

[2]

Rudbeckia purpurea. Purple Rudbeckia.

Class and Order.

Syngenesia Polygamia Frustranea.

Generic Character.

Receptaculum paleaceum, conicum. Pappus margine quadri-dentato.
Calyx duplici ordine squamarum.

Specific Character and Synonyms.

RUDBECKIA *purpurea* foliis lanceolato-ovatis alternis indivisis, radii petalis bifidis. *Linn. Syst. Vegetab. p.* 651. *Sp. Pl. p.* 1280.

DRACUNCULUS virginianus latifolius, petalis florum longissimis purpurascentibus. *Moris. Hist. 3. p.* 42. *f.* 6. *t.* 9. *f.* 1.

This species differs from the other plants of the genus, in the colour of its outermost petals, which are long, narrow, purple, and pendulous, and not unaptly resemble small pieces of red tape. Notwithstanding it is a native of the warm climates Carolina and Virginia, it succeeds very well with us in an open border: but, as Mr. Miller very justly observes, it will always be prudent to shelter two or three plants under a common hot-bed frame in winter, to preserve the kind, because in very severe winters, those in the open air are sometimes killed. It flowers in July. As it rarely ripens its seeds with us, the only mode of propagating it, is by parting the roots; but in that way the plant does not admit of much increase.

[Pg 010]

[3]

Helleborus hyemalis. Winter Hellebore, or Aconite.

Class and Order.

Polyandria Polygynia [A].

Generic Character.

Calyx 0. Petala 5 sive plura. Nectaria bilabiata, tubulata. Capsulæ polyspermæ erectiusculæ.

Specific Character and Synonyms.

HELLEBORUS *hyemalis* flore folio infidente. *Linn. Syst. Vegetab. p.* 431. *Sp. Pl. p.* 783.

ACONITUM unifolium bulbosum. *Bauh. Pin.* 183.

The Winter's Wolfesbane. *Park. Parad. p.* 214.

Grows wild in Lombardy, Italy, and Austria, affects mountainous situations, flowers with us in February, and hence is liable to be cut off by severe frosts. "Is propagated by offsets, which the roots send out in plenty. These roots may be taken up and transplanted any time after their leaves decay, which is generally by the beginning of June till October, when they will begin to put out new fibres; but as the roots are small and nearly the colour of the ground, so if care is not taken to search for them, many of the roots will be left in the ground. These roots should be planted in small clusters, otherwise they will not make a good appearance, for single flowers scattered about the borders of these small kinds are scarce seen at a distance; but when these and the Snowdrops are alternately planted in

bunches, they will have a good effect, as they flower at the same time, and are much of a size." *Millers Gard. Dict.*

FOOTNOTE

[A] Most of the Hellebores vary greatly in the number of their pistils, which in general are too few to justify the placing those plants in the order Polygynia.

[Pg 011]

[Pg 012]

[4]

Cyclamen Coum. Round-leav'd Cyclamen.

Class and Order.

Pentandria Monogynia.

Generic Character.

Corolla rotata, reflexa, tubo brevissimo fauce prominente. Bacca tecta capsula.

Specific Character and Synonyms.

CYCLAMEN *Coum* foliis orbiculatis planis, pediculis brevibus, floribus minoribus. *Miller's Dict.*

CYCLAMEN hyemale orbiculatis foliis inferius rubentibus purpurascente flore; Coum Herbariorum. *Hort. reg. Paris. Herm. Cat.*

CYCLAMEN orbiculato folio inferne purpurascente. *Bauh. Pin. p.* 307.

The common round-leav'd Sowebread. *Park. Parad. p.* 198.

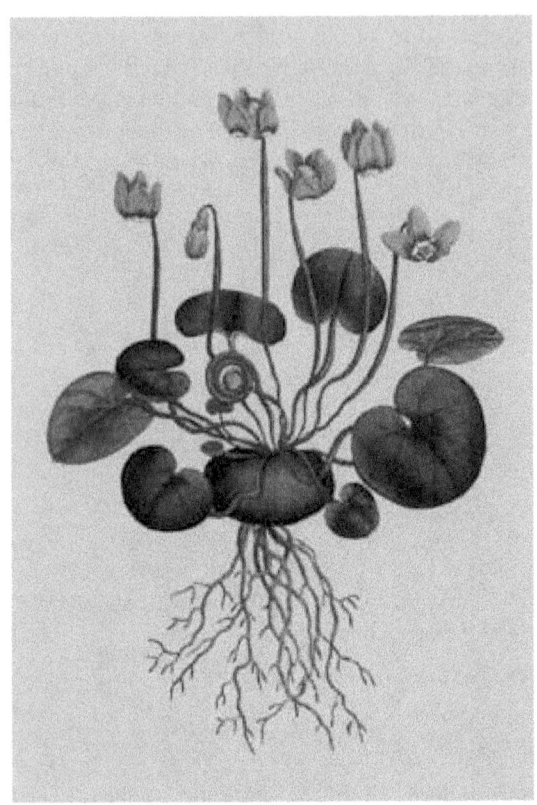

Grows wild in many parts of Italy and Germany, and is sometimes found with white flowers; if the season be mild, or the plants sheltered from the inclemency of the weather, this species will flower as early as February, or much earlier by artificial heat.

As it grows naturally in woods and shady places, it will thrive best in a mixture of bog-earth and loam placed in a north border; if planted in the open border, it will require to be covered with a hand-glass during winter, and in the spring, when in bloom; the more usual method with gardeners is to preserve them in pots in a common hot-bed frame, the advantage of this method is that they may, at any time, be removed to decorate the parlour or the study.

The plants of this genus admit of but little increase by their roots; the best method of propagating them is by seed, which should be sown soon after they are ripe in boxes or pots, and covered about half an inch deep, placing them where they may have only the morning-sun, till the beginning of September, when they may be removed to a warmer exposure.

[Pg 013]

[Pg 014]

[Pg 015]

[5]

Erythronium Dens Canis. Dogs-Tooth, or Dogs-Tooth Violet.

Class and Order.

Hexandria Monogynia.

Generic Character.

Corolla 6-petala, campanulata: Nectario tuberculis 2-petalorum alternorum basi adnatis.

Specific Character and Synonyms.

ERYTHRONIUM *Dens Canis. Lin. Syst. Vegetab. p.* 269. *Sp. Pl. p.* 437.

Dens Canis latiore rotundioreque folio. *Bauh. Pin.* 87.

Dogs-Tooth with a pale purple flower. *Park. Parad. p.* 194.

Of this genus Mr. Miller makes two species; Linnæus, perhaps with more propriety, only one, for breadth of leaves or colour of flowers can scarcely be considered as sufficient to constitute a specific difference.

It is found in the gardens with purple flowers of two different tints, also with white and yellow blossoms, grows naturally in Hungary and some parts of Italy, and blows in the open border at the beginning of April.

"They are propagated by offsets from their roots. They love a shady situation and a gentle loamy soil, but should not be too often removed. They may be transplanted any time after the beginning of June, when their leaves will be quite decayed, till the middle of

September; but the roots should not be kept very long out of the ground, for if they shrink it will often cause them to rot. The roots of these flowers should not be planted scattering in the borders of the flower-garden, but in patches near each other, where they will make a good appearance." *Miller's Gard. Dict.*

[Pg 016]

[Pg 017]

[6]

Narcissus Minor. Least Daffodil.

Class and Order.

Hexandria Monogynia.

Generic Character.

Petala 6, æqualia: Nectario infundibuliformi, 1-phyllo. Stamina intra nectarium.

Specific Character and Synonyms.

NARCISSUS *minor* spatha uniflora, nectario obconico erecto crispo sexfido æquante petala lanceolata. *Lin. Sp. Pl. p.* 415. *Syst. Vegetab. p.* 262.

NARCISSUS parvus totus luteus. *Bauhin. Pin.* 53.

The least Spanish yellow bastard Daffodil. *Park. Parad. p.* 105.

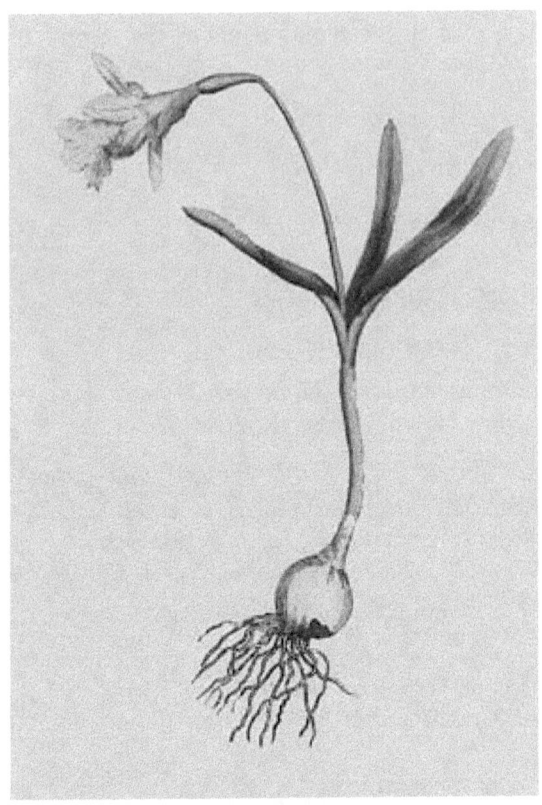

We are not a little surprised that Mr. Miller should have taken no notice of the present species, as it must have been in the English gardens long before his time, being mentioned by Parkinson in his Garden of pleasant Flowers: it is nearly related to the *Pseudo-Narcissus*, but differs from it in many particulars except size, *vid. Lin. Sp. Pl.* and Parkinson above quoted.

Though its blossoms are not so large as those of the other species, yet when the roots are planted in a cluster, they make a very pretty shew, and have this advantage, that they flower somewhat earlier than any of the others.

Like the common Daffodil it propagates very fast by the roots, and will thrive in almost any soil or situation.

Though a native of Spain, it is seldom injured by the severity of our climate.

[Pg 018]

[7]

Cynoglossum Omphalodes. Blue Navelwort.

Class and Order.

Pentandria Monogynia.

Generic Character.

Corolla infundibuliformis, fauce clausa fornicibus. Semina depressa interiore tantum latere stylo affixa.

Specific Character and Synonyms.

CYNOGLOSSUM Omphalodes repens, foliis radicalibus cordatis [B], *Lin. Sp. Pl. p.* 193. *Syst. Vegetab. p.* 157. *Scopoli Fl. Carn. p.* 124. *t.* 3.

SYMPHYTUM minus borraginis facie. *Bauh. Pin.* 259.

BORAGO minor verna repens, folio lævi. *Moris. hist.* 3. *p.* 437. *s.* 11, *t.* 26. *fig.* 3.

A native of Spain, Portugal, and Carniola, and an inhabitant of woods and shady situations, flowers in March and April: in the autumn it puts forth trailing shoots, which take root at the joints, whereby the plant is most plentifully propagated; thrives best under a wall in a North border.

FOOTNOTE

[B] "Stolones repunt non caulis florifer, cui folia ovalia, et minime cordata. TOURNEFORTIUS separavit a Symphito, et dixit Omphallodem *pumilam vernam, symphyti folio*, sed bene monet LINNÆUS solam fructus asperitatem aut glabritiem, non sufficere ad novum genus construendum." *Scopoli Fl. Carn. p.* 124.

[Pg 019]
[Pg 020]

[8]

Helleborus Niger. Black Hellebore, or Christmas Rose.

Class and Order.

Polyandria Polygynia.

Generic Character.

Calyx nullus. Petala 5 sive plura. Nectaria bilabiata, tubulata. Capsulæ polyspermæ, erectiusculæ.

Specific Character and Synonyms.

HELLEBORUS niger scapo sub-bifloro subnudo, foliis pedatis. *Lin. Syst. Vegetab. p.* 431. *Sp. Pl. p.* 783.

HELLEBORUS niger flore roseo, *Bauh. Pin.* 186.

The true Black Hellebore, or Christmas flower. *Parkins. Parad. p.* 344.

Gardeners, in former days, not having that profusion of plants to attend to and cultivate, which we can at present boast, appear to have been more solicitous in increasing generally the varieties of the several species; accordingly, we find in the *Paradisus terrestris* of the venerable Parkinson, no less than six varieties of this plant [C], most of which are now strangers to the Nursery Gardens. We may observe, that varieties in general not being so strong as the original plant, are consequently much sooner lost.

The Iris pumila grows wild in many parts of Hungary, affects open and hilly situations, and flowers in our gardens in the month of April; it is a very hardy plant, and will thrive in almost any soil or situation; is propagated by parting its roots in autumn.

FOOTNOTE

[C]

The	lesser	purple	dwarf	Flower-de-luce	with white blossoms,
"	"	"	"	"	straw colour ditto.
"	"	"	"	"	pale blue ditto.
"	"	"	"	"	blush-coloured ditto.
"	"	"	"	"	yellow variable ditto.
"	"	"	"	"	blue variable ditto, and

the purple dwarf Sea Flower-de-luce of the same author, is probably no other than a variety.

[Pg 024]

[10]

Anemone Hepatica. Hepatica, or Noble Liverwort.

Class and Order.

Polyandria Polygynia.

Generic Character.

Calyx nullus. Petala 6. 9. Semina plura.

Specific Character and Synonyms.

ANEMONE Hepatica foliis trilobis integerrimis. *Lin. Syst. Vegetab. p. 424. Sp. Pl. p. 758. Fl. Suec. n. 480.*

TRIFOLIUM hepaticum flore simplici et pleno. *Bauh. Pin.* 339.

Red Hepatica or noble Liverwort. *Park. Parad. p.* 226.

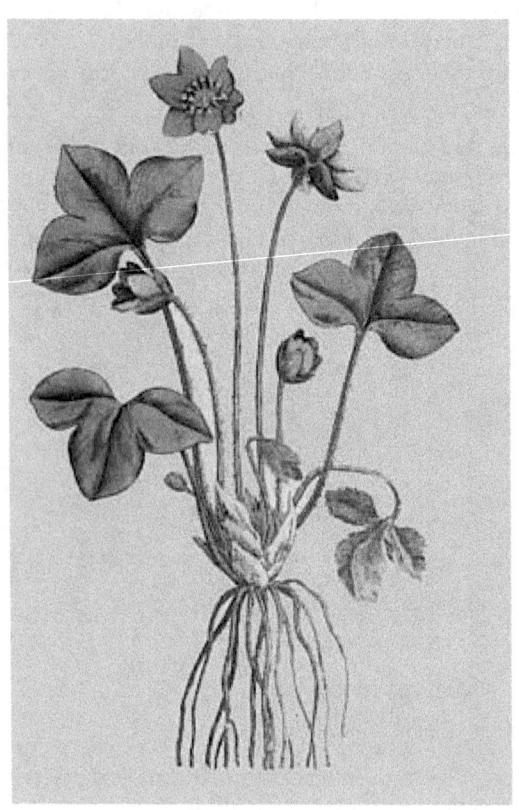

Dillenius, Miller, and some other authors, make a distinct genus of the *Hepatica*: Linnæus unites it with the *Anemone*, observing, that though it differs from the *Anemone* in having a calyx, yet that calyx is at some distance from the flower, and partakes more of the Nature of an Involucrum, which is not uncommon to the Anemonies.

The Hepaticas, as Parkinson observes, flower soon after the winter Hellebore, "and making their pride appear in winter, are the more welcome early guests."

It is found wild in its single state, with red, blue, and white flowers, in the woods and shady mountains of Sweden, Germany, and Italy; the red variety with double flowers is the one most commonly cultivated in our gardens; the double blue is also not unfrequent; the single white is less common; and the double white Miller never saw, yet admits that it may exist spontaneously, or be produced from seed: Parkinson mentions a white variety with red threads or stamina.

According to Miller, this plant delights in a loamy soil, and in an eastern position where it may have only the morning sun: the single sorts are easily raised from seed; the double, increased by parting the roots, which ought to be done in March when they are in bloom; they should not be divided into very small heads: these plants, if often removed and parted, are apt to die, but left undisturbed for many years, they will thrive exceedingly, and become very large roots.

[Pg 025]

[Pg 026]

[Pg 027]

[11]

Erica herbacea. Herbaceous Heath.

Class and Order.

Octandria Monogynia.

Generic Character.

Calyx 4-phyllus. Corolla 4-fida. Filamenta receptaculo inserta. Antheræ bifidæ. Capsula 4-locularis.

Specific Character and Synonyms.

ERICA *herbacea* antheris muticis exsertis, corollis oblongis, stylo exserto, foliis quaternis, floribus secundis, *Lin. Syst. Vegetab. p.* 306. *carnea Sp. Pl. ed.* 3. *p.* 504.

ERICA *carnea. Jacq. Fl. Austr. v.* 1. *tab.* 32

ERICA procumbens herbacea. *Bauh. Pin. p.* 486.

Since the days of Mr. Miller, who, with all his imperfections, has contributed more to the advancement of practical gardening than any individual whatever, our gardens, but more especially our green-houses, have received some of their highest ornaments from the introduction of a great number of most beautiful Heaths: the present plant, though a native of the Alps and mountainous parts of Germany, is of modern introduction here, what renders it particularly acceptable, is its hardiness and early flowering; its blossoms are formed in the autumn, continue of a pale green colour during the winter, and expand in the spring, flowering as early as March,

especially if kept in a green-house, or in a common hot-bed frame, which is the more usual practice.

It may be propagated by seeds or cuttings, the latter is the most ready way of increasing this and most of the other species of the genus: when the cuttings have struck root, they should be planted in a mixture of fresh loam and bog earth, either in the open border, under a wall, or in pots.

The name of *herbacea*, which Linnæus has given to this plant, is not very characteristic, but it should be observed, that Linnæus in this, as in many other instances, has only adopted the name of some older botanist; and it should also be remembered, that in genera, where the species are very numerous, it is no easy matter to give names to all of them that shall be perfectly expressive.

This species does not appear to us to be specifically different from the *mediterranea*.

[Pg 028]

[Pg 029]

[12]

Dodecatheon Meadia. Mead's Dodecatheon, or American Cowslip.

Class and Order.

Pentandria Monogynia.

Generic Character.

Corolla rotata, reflexa. *Stamina* tubo insidentia. *Capsula* unilocularis, oblonga.

Specific Character and Synonyms.

DODECATHEON *Meadia. Lin. Syst. Vegetab. p.* 163. *Sp. Plant. p.* 163.

MEADIA *Catesb. Car.* 3. *p.* 1. *t.* 1. *Trew. Ehret. t.* 12.

AURICULA ursi virginiana floribus boraginis instar rostratis, cyclaminum more reflexis. *Pluk. alm.* 62. *t.* 79. *f.* 6.

This plant grows spontaneously in Virginia and other parts of North America, from whence, as Miller informs us, it was sent by Mr. Banister to Dr. Compton, Lord Bishop of London, in whose curious garden he first saw it growing in the year 1709.

It is figured by Mr. Catesby, in his Natural History of Carolina, among the natural productions of that country, who bestowed on it the name of *Meadia*, in honour of the late Dr. Mead, a name which Linnæus has not thought proper to adopt as a generic, though he has as a trivial one.

"It flowers the beginning of May, and the seeds ripen in July, soon after which the stalks and leaves decay, so that the roots remain inactive till the following spring.

"It is propagated by offsets, which the roots put out freely when they are in a loose moist soil and a shady situation; the best time to remove the roots, and take away the offsets, is in August, after the leaves and stalks are decayed, that they may be fixed well in their new situation before the frost comes on. It may also be propagated by seeds, which the [Pg 030]plants generally produce in plenty; these should be sown in autumn, soon after they are ripe, either in a shady moist border, or in pots, which should be placed in the shade; in the spring, the plants will come up, and must then be kept clean from weeds; and, if the season proves dry, they must be frequently refreshed with water: nor should they be exposed to the sun; for while the plants are young, they are very impatient of heat, so that I have known great numbers of them destroyed in two or three days, which were growing to the full sun. These young plants should not be transplanted till the leaves are decayed, then they may be carefully taken up and planted in a shady border, where the soil is loose and moist, at about eight inches distance from each other, which will be room enough for them to grow one year, by which time they will be strong enough to produce flowers, so may then be transplanted into some shady borders in the flower-garden, where they will appear very ornamental during the continuance of their flowers." *Miller's Gard. Dict.*

[Pg 031]

[Pg 032]

[13]

Coronilla glauca. Sea-green, or Day-smelling Coronilla.

Class and Order.

Diadelphia Decandria.

Generic Character.

Calyx bilabiatus: 2/3: dentibus superioribus connatis. *Vexillum* vix alis longius.
Legumen isthmis interceptum.

Specific Character and Synonyms.

CORONILLA *glauca* fruticosa, foliolis septenis, obtusissimis, stipulis lanceolatis. *Linn. Syst. Vegetab. p.* 557. *Sp. Pl.* 1047.

CORONILLA maritima, glauco folio. *Tournef. inst.* 650.

COLUTEA scorpioides maritima, glauco folio. *Bauh. Pin.* 397. *prodr.* 157.

This charming shrub, which is almost perpetually in blossom, and admirably adapted for nosegays, is a native of the south of France, and a constant ornament to our green-houses.

Linnæus has observed, that the flowers, which in the day time are remarkably fragrant, in the night are almost without scent.

"It is propagated by sowing the seeds in the spring, either upon a gentle hot-bed, or on a warm border of light earth: when the plants are come up about two inches high, they should be transplanted either into pots, or into a bed of fresh earth, at about four or five inches distance every way, where they may remain until they have obtained strength enough to plant out for good, which should be either in pots filled with good fresh earth, or in a warm situated border, in which, if the winter is not too severe, they will abide very well, provided they are in a dry soil." *Miller's Gard. Dict.*

[Pg 033]

[Pg 034]

[Pg 035]

[14]

Primula villosa. Mountain Primula.

Class and Order.

Pentandria Monogynia.

Generic Character.

Involucrum umbellulæ. *Corollæ* tubus cylindricus: ore patulo.

Specific Character and Synonyms.

PRIMULA *villosa* foliis obovatis dentatis villosis, scapo brevissimo multifloro.

PRIMULA *villosa. Jacquin Fl. Austr. app. t.* 27.

Mr. Miller, in the Sixth Edition of the Abridgment of his Gardener's Dictionary, mentions only four Primulas, exclusive of the Auricula, the two first of which are named erroneously, and of the two last not a syllable is said either as to their place of growth or culture.

The plant here figured, has been introduced pretty generally into the Nursery-Gardens in the neighboured of London within these few years: Mr. Salisbury informs me, that a variety of this plant with white flowers, brought originally from the Alps of Switzerland, has for many years been cultivated in a garden in Yorkshire.

It is not noticed by Linnæus: Professor Jacquin, in his Flora Austriaca, has figured and described a Primula, which, though not

agreeing so minutely as could be wished with the one we have figured, is nevertheless considered by some of the first Botanists in this country as the same species; he gives it the name of *villosa*, which we adopt, though with us it is so slightly villous as scarcely to deserve that epithet.

It varies in the brilliancy of its colours, flowers in April, and will succeed with the method of culture recommended for the Round-Leaved Cyclamen.

[Pg 036]

[Pg 037]

[15]

Narcissus Jonquilla. Common Jonquil.

Class and Order.

Hexandria Monogynia.

Generic Character.

Petala sex. *Nectario* infundibuliformi, monophyllo.

Stamina intra nectarium.

Specific Character and Synonyms.

NARCISSUS *Jonquilla* spatha multiflora, nectario hemisphærico crenato, breviore petalis, foliis semiteretibus. *Lin. Spec. Pl. p.* 417.

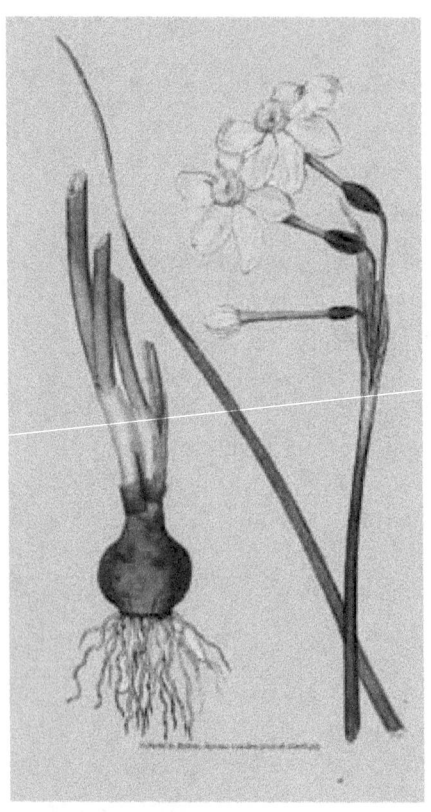

The fragrant Jonquil is a native of Spain, flowers in the open ground, about the latter-end of April, or beginning of May, and will thrive in almost any soil or situation, but prefers, as most bulbs do, a fresh loamy earth; indeed such a soil is favourable to the growth of most plants, as being exempt from a variety of subterraneous insects, which are apt to infest ground which has been long cultivated.

It is found in the gardens with double flowers.

Our plant accords exactly with the description of Linnæus, above quoted, but must be carefully distinguished from some others very similar to it.

[Pg 038]

[16]

Iris variegata. Variegated Iris.

Class and Order.

Triandria Monogynia.

Generic Character.

Corolla 6-partita; *Petalis* alternis, reflexis. *Stigmata* petaliformia.

Specific Character and Synonyms.

> IRIS *variegata* corollis barbatis, caule subfolioso longitudine foliorum multifloro. *Linn. Spec. Pl. p.* 56.

IRIS latifolia pannonica, colore multiplici. *Bauh. Pin.* 31.

The yellow variable Flower-de-Luce. *Parkinson Parad. p.* 182.

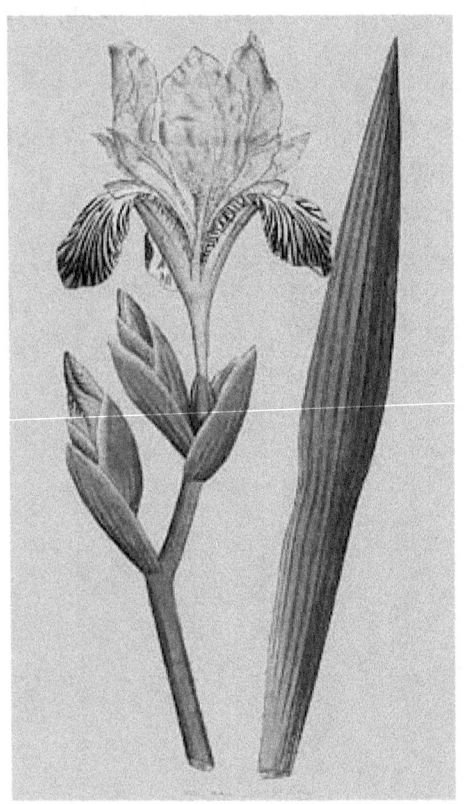

This species of Iris, inferior to few in point of beauty, is a native of the hilly pastures of Hungary, and flowers in our gardens in the month of May, and beginning of June. It is a hardy perennial, requires no particular treatment, and may be easily propagated by parting its roots in Autumn.

[Pg 039]

[Pg 040]

[17]

Cactus flagelliformis. Creeping Cereus.

Class and Order.

Icosandria Monogynia.

Generic Character.

Calyx 1-phyllus, superus, imbricatus. *Corolla* multiplex. *Bacca* 1-locularis, polysperma.

Specific Character.

CACTUS *flagelliformis* repens decemangularis. *Linn. Syst. Vegetab. ed.* 14 *p.* 460.

CEREUS *flagelliformis. Miller's Gard. Dict. ed.* 6. 4*to.*

Grows spontaneously in South-America, and the West-Indies, flowers in our dry stoves early in June, is tolerably hardy, and will thrive even in a common green-house, that has a flue to keep out the severe frosts.

It is superior to all its congeners in the brilliancy of its colour, nor are its blossoms so fugacious as many of the other species.

No plant is more easily propagated by cuttings; these Miller recommends to be laid by in a dry place for a fortnight, or three weeks, then to be planted in pots, filled with a mixture of loam and lime rubbish, having some stones laid in the bottom of the pot to drain off the moisture, and afterwards plunged into a gentle hot-bed of

Tanners bark, to facilitate their rooting, giving them once a week a gentle watering: this business to be done the beginning of July.

It is seldom that this plant perfects its seeds in this country: Miller relates that it has borne fruit in Chelsea gardens.

[Pg 041]

[Pg 042]

[Pg 043]

[18]

Geranium Reichardi. Dwarf Geranium.

Class and Order.

Monadelphia Decandria.

General Character.

Monogynia. Stigmata 5. Fructus rostratus, 5-coccus.

Specific Character and Synonyms.

GERANIUM *Reichardi* scapis unifloris, floribus pentandris, foliis subreniformibus inciso-crenatis.

GERANIUM *Reichardi* scapis unifloris, foliis plerisque oblongis trilobis vel quinquelobis inciso-crenatis. *Linn. Syst. Vegetab. ed. Murr.* 14. *p.* 618.

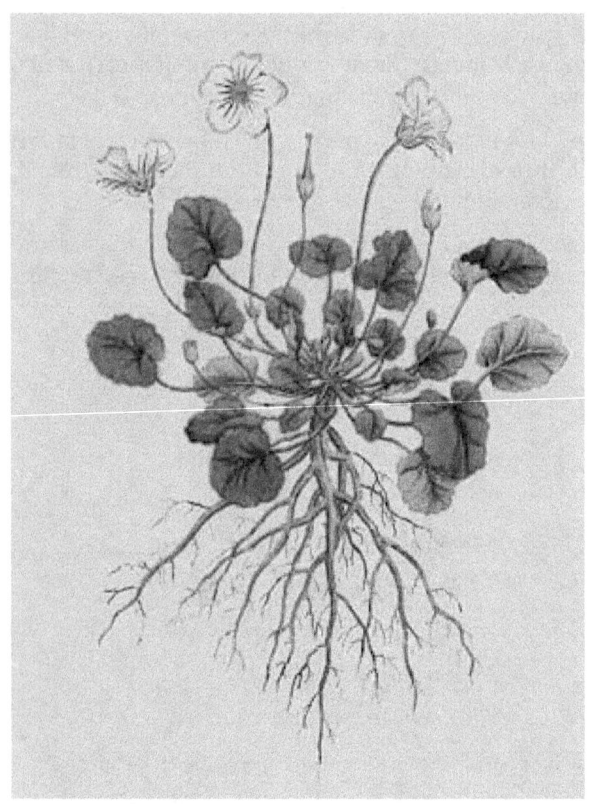

This species of Geranium, so strikingly different from all others at present cultivated in our gardens, has been known for several years to the Nursery-men in the neighbourhood of London, by the name of *acaule*, a name we should gladly have retained, had not Professor Murray described it in the 14th edition of Linnæus's *Systema Vegetabilium*, under the name of *Reichardi*, a name he was disposed to give it in compliment to a French gentleman, who first discovered it in the island of Minorca, and introduced it into the gardens of France.

Linnæus describes many of the Geraniums, as having only five antheræ, though several of those he thus describes have to our certain knowledge ten, the five lowermost of which shedding their pollen first, often drop off, and leave the filaments apparently bar-

ren: but in this species (with us at least) there never are more than five, but betwixt each stamen, there is a broad pointed barren filament or squamula, scarcely to be distinguished by the naked eye.

The usual and best practice is to make a green-house plant of this species, though it has been known to remain in the open ground, during a mild winter, unhurt.

It continues to have a succession of blossoms during the greatest part of the summer, and may be propagated either by seed or parting its roots.

[Pg 044]

[19]

Hemerocallis Flava. Yellow Day-lily.

Class and Order.

Hexandria Monogynia.

Generic Character.

Corolla campanulata, tubus cylindraceus.

Stamina declinata.

Specific Character and Synonyms.

HEMEROCALLIS *flava* foliis lineari-subulatis carinatis, corollis flavis. *Linn. Syst. Veg. ed.* 14. *p.* 339.

LILIUM luteum, asphodeli radice. *Bauh. Pin.* 80.

The Yellow Day-Lily. *Parkins. Parad. p.* 148.

This Genus has been called *Hemerocallis*, in English, *Day-Lily*, from the short duration of its blossoms, but these are not quite so fugacious in this species as in the *fulva*.

It very rarely happens that Linnæus, in his specific character of a plant, has recourse to colour, he has however in this instance; but this seems to arise from his considering them rather as varieties, than species. To us they appear to be perfectly distinct, and in addition to several other characters, the flava is distinguished by the fragrance of its blossoms.

This species is an inhabitant of Hungary and Siberia, and consequently bears our climate exceedingly well; it requires a moist soil,

and a situation somewhat shady, and is easily propagated by parting its roots in autumn.

[Pg 045]
[Pg 046]
[Pg 047]

[20]

Geranium Peltatum. Ivy-Leaved Geranium.

Class and Order.

Monadelphia Decandria.

Generic Character.

Monogyna. *Stigmata* quinque. *Fructus* rostratus. 5-coccus.

Specific Character.

GERANIUM *peltatum* calycibus monophyllis, foliis quinquelobis integerrimis glabris subpeltatis, caule fruticoso. *Linn. Syst. Vegetab.* ed. 14. *p.* 613.

GERANIUM africanum, foliis inferioribus asari, superioribus staphidisagriæ maculatis splendentibus et acetosæ sapore. *Comm. Præl.* 52. *t.* 2.

A native of Africa, as are most of our shewy Geraniums, is not so tender as many others, and may be propagated very readily from cuttings.

A leaf, having its foot-stalk inserted into the disk or middle part of it, or near it, is called by Linnæus, peltatum, hence the Latin trivial name of this plant. It may be observed, however, that some of the leaves have this character more perfectly than others.

The African Geraniums differ much from the European, in the irregularity of their Petals, but exhibit the character of the Class *Monadelphia* much better than any of our English ones, having their filaments manifestly united into one body; this species has only 7

filaments bearing antheræ, but 3 barren ones may be discovered upon a careful examination, which makes it of the order *Decandria*.

[Pg 048]

[Pg 049]

[21]

Iris Versicolor. Particoloured Iris.

Class and Order.

Triandria Monogynia.

Generic Character.

Corolla 6-petala, inæqualis, petalis alternis geniculato-patentibus. *Stigmata* petaliformia, cucullato-bilabiata. Conf. *Thunb. Dis. de Iride.*

Specific Character and Synonyms.

> IRIS *versicolor* imberbis foliis ensiformibus, scapo tereti flexuoso, germinibus subtrigonis. *Linn. Syst. Vegetab. ed.* 14. *Murr. p.* 90. *Sp. Plant. ed.* 3. *p.* 57.

IRIS Americana versicolor stylo crenato. *Dill. Elth.* 188. 1. 155. *f.* 188.

A native of Virginia, Maryland, and Pensylvania, has a perennial root, is hardy, and will thrive in almost any soil or situation; may be increased by parting its roots in autumn.

Our plant is the *picta* of Miller, and the *versicolor* of Miller is, we believe, the *sibirica* of Linnæus.

This species has, for the most part, a stalk unusually crooked or elbowed, by which it is particularly distinguished. It flowers in June, as do most of this beautiful tribe.

[Pg 050]

[22]

Nigella damascena. Garden Fennel-flower, Love in a mist, Devil in a Bush.

Class and Order.

Polyandria Pentagynia.

Generic Character.

Cal. nullus. *Petala* 5. Nectaria 5. trifida, intra corollam. *Capsulæ* 5 connexæ.

Specific Character and Synonyms.

NIGELLA *damascena* floribus involucro folioso cinctis. *Lin. Syst. Vegetab. ed.* 14. *Murr. p.* 506. *Sp. Pl. p.* 753.

NIGELLA angustifolia, flore majore simplici cæruleo. *Bauh. Pin.* 145.

The great Spanish Nigella. *Park. Parad. p.* 287.

Is an annual, and grows wild among the corn in the southern parts of Europe; varies with white and blue flowers, both single and double.

"May be propagated by sowing their seeds upon a bed of light earth, where they are to remain (for they seldom succeed well if transplanted); therefore, in order to have them intermixed among other annual flowers in the borders of the Flower Garden, the seeds should be sown in patches at proper distances: and when the plants come up, they must be thinned where they grow too close, leaving but three or four of them in each patch, observing also to keep them clear from weeds, which is all the culture they require. In July they will produce their flowers, and their seeds will ripen in August.

"The season for sowing these seeds is in March; but if you sow some of them in August, soon after they are ripe, upon a dry soil and in a warm situation, they will abide through the winter, and flower strong the succeeding year; by sowing of the seeds at different times, they may be continued in beauty most parts of the summer." *Miller's Gard. Dict. ed.* 6. *4to.*

[Pg 051]

[Pg 052]

[Pg 053]

[23]

Tropæolum majus. Greater Indian-Cress, or Nasturtium.

Class and Order.

Octandria Monogynia.

Generic Character.

Calyx 1-phyllus, calcaratus. *Petala* 5 in æqualia. *Baccæ* tres, siccæ.

Specific Character and Synonyms.

TROPÆOLUM *majus* foliis peltatis subquinquelobis, petalis obtusis. *Lin. Syst. Vegetab. ed.* 14. *Murr. p.* 357. *Sp. Pl. p.* 490.

CARDAMINDUM ampliori folio et majori flore. *Grande Capucine Tournef. Inst. p.* 430.

The present plant is a native of Peru, and is said by Linnæus to have been first brought into Europe in the year 1684; it is certainly one of the greatest ornaments the Flower-Garden can boast: it varies in colour, and is also found in the Nurseries with double flowers. The former, as is well known, is propagated by seed; the latter by cuttings, which should be struck on a hot-bed. To have these plants early, they should be raised with other tender annuals; they usually begin to flower in July, and continue blossoming till the approach of winter: the stalks require to be supported, for if left to themselves they trail on the ground, overspread, and destroy the neighbouring plants.

Elizabeth Christina, one of the daughters of Linnæus, is said to have perceived the flowers to emit spontaneously, at certain intervals, sparks like those of electricity, visible only in the dusk of the evening, and which ceased when total darkness came on.

The flowers have the taste of water-cress, with a degree of sweetness, which that plant does not possess, more particularly resident in the spur of the calyx or nectary; hence are sometimes used in sallads, and hence the plant acquires its name of *Nasturtium*.

[Pg 054]

[Pg 055]

[24]

Agrostemma coronaria. Rose Cockle, or Campion.

Class and Order.

Decandria Pentagynia.

Generic Character.

Calyx 1-phyllus, coriaceus. *Petala* 5 unguiculata: limbo obtuso, indiviso.
Caps. 1-locularis.

Specific Character and Synonyms.

AGROSTEMMA *coronaria* tomentosa, foliis ovato-lanceolatis, petalis emarginatis coronatis serratis. *Lin. Syst. Vegetab. ed.* 14. *Murr. p.* 435. *Sp. Pl. p.*

LYCHNIS coronaria dioscoridis sativa. *Bauh. Pin.* 203. The single red Rose Campion. *Parkins. Parad. p.* 252.

Grows spontaneously in Italy and Siberia; Linnæus informs us that the blossom is naturally white, with red in the middle.

"The single Rose Campion has been long an inhabitant of the English gardens, where, by its seeds having scattered, it is become a kind of weed. There are three varieties of this plant, one with deep red, another with flesh-coloured, and a third with white flowers, but these are of small esteem, for the double Rose Campion being a finer flower, has turned the others out of most fine gardens. The single sorts propagate fast enough by the seeds, the sort with double flowers never produces any, so is only propagated by parting of the roots; the best time for this is in autumn, after their flowers are past; in doing of this, every head which can be slipped off with

roots should be parted; these should be planted in a border of fresh undunged earth, at the distance of six inches, observing to water them gently until they have taken root, after which they will require no more, for much wet is injurious to them, as is also dung. After the heads are well rooted, they should be planted into the borders of the Flower-Garden, where they will be very ornamental during the times of their flowering, which is in July and August." *Miller's Gard. Dict. ed.* 6. 4*to.*

Miller, by mistake, calls this plant *Cælirosa*.

[Pg 056]

[25]

Dianthus chinensis. China or Indian Pink.

Class and Order.

Decandria Digynia.

Generic Character.

Calyx cylindricus, 1-phyllus: basi squamis 4. *Petala* 5, unguiculata. *Capsula* cylindrica, 1-locularis.

Specific Character and Synonyms.

DIANTHUS *chinensis* floribus solitariis, squamis calycinis subulatis patulis, tubum æquantibus, corollis crenatis. *Lin. Syst. Vegetab. p.* 418. *Sp. Pl.* 588.

CARYOPHYLLUS sinensis supinus, leucoji folio, flore unico. *Tournef. act.* 1705. *p.* 348. *f.* 5.

This species, unknown to the older botanists, is a native of China, hence its name of China Pink; but, in the nurseries, it is in general better known by the name of Indian Pink.

Though it cannot boast the agreeable scent of many of its congeners, it eclipses most of them in the brilliancy of its colours; there are few flowers indeed which can boast that richness and variety found among the most improved varieties of this species; and as these are easily obtained from seed, so they are found in most collections, both single and double.

It is little better than an annual, but will sometimes continue two years in a dry soil, which it affects.

Attempts have been made to force it, but, as far as we have learned, with no great success.

[Pg 057]

[Pg 058]

[Pg 059]

[26]

Stapelia variegata. Variegated Stapelia.

Class and Order.

Pentandria Digynia.

Generic Character.

Contorta. *Nectarium* duplici stellula tegente genitalia.

Specific Character and Synonyms.

STAPELIA *variegata* denticulis ramorum patentibus. *Lin. Syst. Vegetab. p.* 260. *Sp. Pl. p.* 316.

ASCLEPIAS aizoides africana. *Bradl. suc.* 3. *p.* 3. *t.* 22.

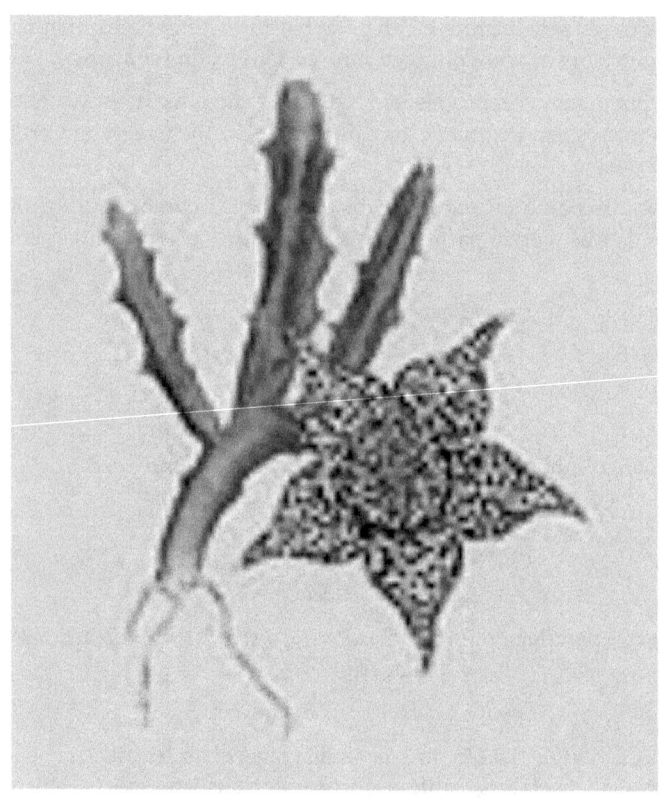

This very singular plant is a native of the Cape of Good Hope, where it grows and flourishes on the rocks with the *Stapelia hirsuta*.

If these plants be kept in a very moderate stove in winter, and in summer placed in an airy glass-case where they may enjoy much free air, but screened from wet and cold, they will thrive and flower very well; for although they will live in the open air in summer, and may be kept through the winter in a good green-house; yet these plants will not flower so well as those managed in the other way. They must have little water given them, especially in winter.

It is very seldom that the *variegata* produces seed-vessels in this country; Miller observes, in upwards of forty years that he cultivat-

ed it, he never saw it produce its pods but three times, and then on such plants only as were plunged into the tan-bed in the stove.

This plant may be propagated without seeds, as it grows fast enough from slips; treatment the same as that of the Creeping Cereus, which see.

It takes its name of *Stapelia* from *Stapel*, a Dutchman, author of some botanical works, particularly a Description of Theophrastus's plants.

[Pg 060]

[Pg 061]

[27]

Convolvulus tricolor. Small Convolvulus or Bindweed.

Class and Order.

Pentandria Monogynia.

Generic Character.

Corolla campanulata, plicata. *Stigmata* 2. *Capsula* 2-locularis: loculis dispermis.

Specific Character and Synonyms.

CONVOLVULUS *tricolor* foliis lanceolato ovatis glabris, caule declinato, floribus solitariis. *Lin. Syst. Vegetab. p.* 203. *Sp. Pl. p.* 225.

CONVOLVULUS peregrinus cæruleus, folio oblongo. *Bauh. Pin.* 295. Flore triplici colore insignito. *Moris. hist.* 2. *p.* 17. *s.* 1. *t.* 4. *f.* 4.

The Spanish Small Blew Bindeweede. *Parkins. Parad. p.* 4.

This species has usually been called *Convolvulus minor* by gardeners, by way of distinguishing it from the *Convolvulus purpureus*, to which they have given the name of *major*. It is a very pretty annual; a native of Spain, Portugal, and Sicily, and very commonly cultivated in gardens.

The most usual colours of its blossoms are blue, white, and yellow, whence its name of *tricolor*; but there is a variety of it with white, and another with striped blossoms.

The whole plant with us is in general hairy, hence it does not well accord with Linnæus's description. It is propagated by seeds, which should be sown on the flower-borders in the spring, where the

plants are to remain: they require no other care than to be thinned and weeded.

[Pg 062]

[28]

Passiflora cœrulea. Common Passion-Flower.

Class and order.

Gynandria Hexandria.

Generic Character.

Trigyna. *Cal.* 5-phyllus. *Petala* 5. *Nectarium* corona. *Bacca* pedicellata.

Specific Character and Synonyms.

PASSIFLORA *cœrulea* foliis palmatis integerrimis. *Lin. Syst. Vegetab. p.* 823. *Sp. Pl. p.* 1360.

GRANADILLA polyphyllos, fructu ovato. *Tourn. inst.* 241.

FLOS PASSIONIS major pentaphyllus. *Sloan. Jam.* 104. *hist.* 1. *p.* 229.

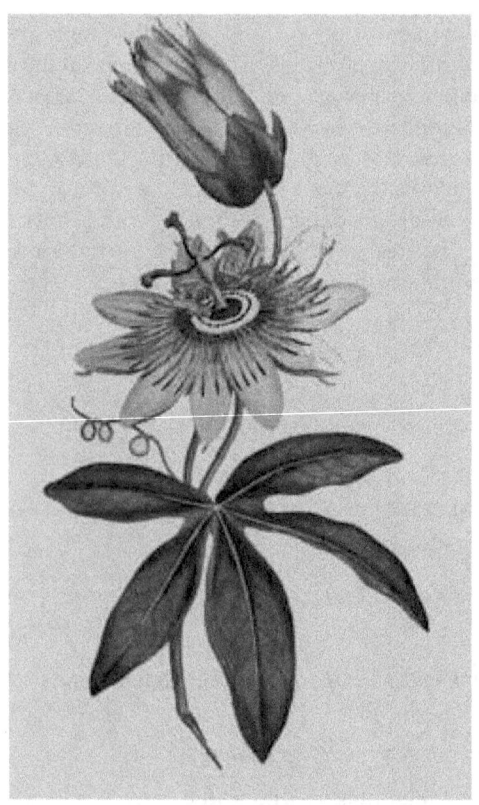

The Passion-Flower first introduced into this country was the *incarnata* of Linnæus, a native of Virginia, and figured by Parkinson in his *Paradisus Terrestris*, who there styles it the surpassing delight of all flowers: the present species, which, from its great beauty and superior hardiness, is now by far the most common, is of more modern introduction; and, though a native of the Brasils, seldom suffers from the severity of our climate; flowering plentifully during most of the summer months, if trained to a wall with a southern aspect, and, in such situations, frequently producing ripe fruit, of the size and form of a large olive, of a pale orange colour.

This most elegant plant may be propagated by seeds, layers, or cuttings; foreign seeds are most to be depended on; they are to be

sown in the spring, on a moderate hot-bed, and when the plants are grown to the height of two or three inches, they are to be carefully taken up, and each planted in a separate small pot, filled with good loam, then plunged into a moderate hot-bed, to forward their taking new root; after which they should be gradually inured to the common air: the younger the plants the more shelter they require, and if ever so old or strong, they are in danger from severe frosts. The layers and cuttings are to be treated in the common way, but seedling plants, if they can be obtained, are on many accounts to be preferred.

[Pg 063]

[Pg 064]

[29]

Reseda odorata. Sweet-scented Reseda or Mignonette.

Class and Order.

Dodecandria Trigynia.

Generic Character.

Cal. 1-phyllus, partitus. *Petala* laciniata. *Caps.* ore dehiscens, 1-locularis.

Specific Character and Synonyms.

RESEDA *odorata* foliis integris trilobisque, calycibus florem æquantibus. *Lin. Syst. Vegetab.* p. 449.

RESEDA foliis integris trilobisque, floribus tetragynis. *Mill. Dict. t.* 217.

Mignonette grows naturally in Egypt, it was unknown to the older Botanists; Miller says he received the seeds of it from Dr. Adrian Van Royen, Professor of Botany at Leyden, so that it is rather a modern inhabitant of our gardens.

The luxury of the pleasure-garden is greatly heightened by the delightful odour which this plant diffuses; and as it is most readily cultivated in pots, its fragrance may be conveyed to the parlour of the recluse, or the chamber of the valetudinarian; its perfume, though not so refreshing perhaps as that of the Sweet-Briar, is not apt to offend on continuance the most delicate olfactories.

Being an annual it requires to be raised yearly from seed; when once introduced on a warm dry border it will continue to sow itself,

and grow very luxuriantly, flowering from June to the commencement of winter; but as it is desirable to have it as early as possible in the spring, the best way is either to sow the seed in pots in autumn, securing them through the winter in frames, or in a greenhouse, or to raise the seeds early on a gentle hot bed, thinning the plants if they require it, so as to have only two or three in a pot.

[Pg 065]

[Pg 066]

[Pg 067]

[30]

Lilium chalcedonicum. Chalcedonian Lily.

Class and Order.

Hexandria Monogynia.

Generic Character.

Cor. 6-petala, campanulata: *linea* longitudinali nectarifera. *Caps.* valvulis pilo cancellato connexis.

Specific Character and Synonyms.

LILIUM *chalcedonicum* foliis sparsis lanceolatis, floribus reflexis, corollis revolutis. *Lin. Syst. Vegetab. p.* 324.

LILIUM byzantium miniatum. *Bauh. Pin.* 78.

The Red Martagon of Constantinople. *Park. Parad. p.* 34.

This species is best known in the nurseries by the name of the *Scarlet Martagon*; but as it is not the Martagon of Linnæus, to avoid confusion it will be most proper to adhere to the name which Linnæus has given it.

It is a native not only of Persia, but of Hungary; Professor Jacquin, who has figured it in his most excellent *Flora Austriaca*, describes it as growing betwixt Carniola and Carinthia, and other parts of Hungary, but always on the tops of the largest mountains.

It varies in the number of its flowers, from one to six, and the colour in some is found of a blood red.

Authors differ in their ideas of its smell: Jacquin describing it as disagreeble, while Scopoli compares it to that of an orange.

It flowers in June and July; and is propagated by offsets, which it produces pretty freely, and which will grow in almost any soil or situation.

The best time for removing the roots is soon after the leaves are decayed, before they have begun to shoot.

[Pg 068]

[31]

Jasminum officinale. Common Jasmine or Jessamine.

Class and Order.

Diandria Monogynia.

Generic Character.

Cor. 5-fida. *Bacca* dicocca. *Sem.* arillata. *Antheræ* intra tubum.

Specific Character and Synonyms.

JASMINUM *officinale* foliis oppositis; foliolis distinctis. *Lin. Syst. Vegetab. p.* 56.

JASMINUM vulgatius flore albo. *Bauh. Pin.* 397.

Jasmine or Gesmine. *Park. Parad. p.* 406.

There is an elegance in the Jasmine which added to its fragrance renders it an object of universal admiration.

"It grows naturally at Malabar, and in several parts of India, yet has been long inured to our climate, so as to thrive and flower extremely well, but never produces any fruit in England. It is easily propagated by laying down the branches, which will take root in one year, and may then be cut from the old plant, and planted where they are designed to remain: it may also be propagated by cuttings, which should be planted early in the autumn, and guarded against the effects of severe frosts.

"When these plants are removed, they should be planted either against some wall, pale, or other fence, where the flexible branches

may be supported. These plants should be permitted to grow rude in the summer, otherwise there will be no flowers; but after the summer is past, the luxuriant shoots should be pruned off, and the others must be nailed to the support.

"There are two varieties of this with variegated leaves, one with white, the other with yellow stripes, but the latter is the most common: these are propagated by budding them on the plain Jasmine; they require to be planted in a warm situation, especially the white-striped, for they are much more tender than the plain, and in very severe winters their branches should be covered with mats or straw to prevent their being killed." *Miller's Gard. Dict.*

[Pg 069]
[Pg 070]
[Pg 071]

[32]

Mesembryanthemum dolabriforme. Hatchet-leav'd Fig-Marigold.

Class and Order.

Icosandria Pentagynia.

Generic Character.

Cal. 5-fidus. *Petala* numerosa, linearia. *Caps.* carnosa infera, polysperma.

Specific Character and Synonyms.

MESEMBRYANTHEMUM *dolabriforme* acaule, foliis dolabriformibus punctatis. *Lin. Syst. Veg. p.* 470.

FICOIDES capensis humilis, foliis cornua cervi referentibus, petalis luteis noctiflora, *Bradl. suc.* 1. *p.* 11. *t.* 10. *Dillen Hort. Elth. t.* 191. *f.* 237.

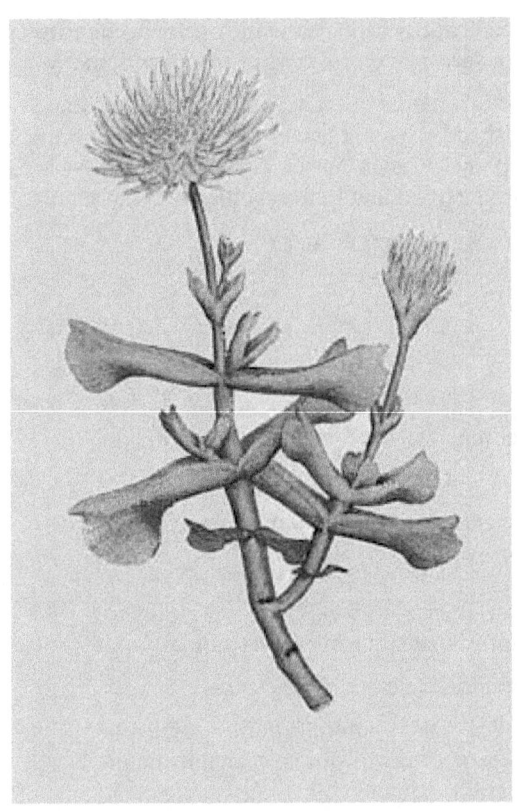

Though many Latin names of plants, as *Geranium*, *Hepatica*, *Convolvulus*, &c. are more familiar to the ear, and more generally used than their English ones, yet *Mesembryanthemum* though used by some, appears too long to be generally adopted, its English name of *Fig-marigold* is doubtless to be preferred.

The Fig-marigolds are a very numerous tribe, chiefly inhabitants of the Cape of Good Hope; no less than thirty-three species are figured in that inestimable work the *Hortus Elthamensis* of Dillenius. As most of these plants grow readily from slips, or cuttings, and require only the shelter of a common greenhouse, and as they recommend themselves to our notice, either from the extreme singular-

ity of their foliage, the beauty of their flowers, or the peculiarity of their expansion, so they are a favourite class of plants with many.

The present species is a native of the Cape of Good Hope, and is particularly distinguished by having leaves somewhat resembling a hatchet, whence its name; it is as hardy as most, and flowers as freely, but its blossoms fully expand in the evening and night only.

It is very readily propagated by cuttings.

[Pg 072]

[Pg 073]

[33]

Aster tenellus. Bristly-leav'd Aster.

Class and Order.

Syngenesia Polygamia Superflua.

Generic Character.

Recept. nudum. *Pappus* simplex. *Cor.* radii plures 10. *Cal.* imbricati squamæ inferiores patulæ.

Specific Character and Synonyms.

ASTER *tenellus* foliis subfiliformibus aculeato-ciliatis, pedunculis nudis, calycibus hemisphæricis æqualibus. *Lin. Syst. Vegetab. p.* 760.

ASTER parvus æthiopicus, chamæmeli floribus, tamarisci ægyptiaci foliis tenuissime denticulatis. *Pluk. alm.* 56. *t.* 271. *f.* 4. *Raii. Suppl.* 164. *n.* 84.

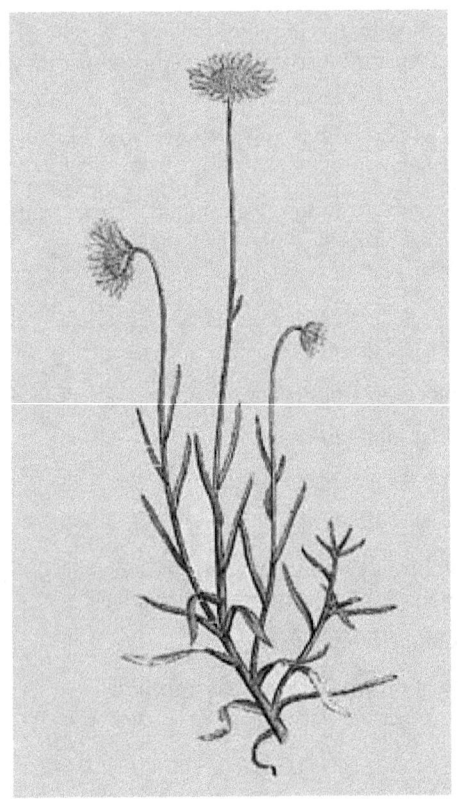

Most of the numerous species of this genus flower about Michaelmas, hence their vulgar name of *Michaelmas-Daisy*; a name exceptionable not only on account of its length, but from its being a compound word. *Aster*, though a Latin term, is now so generally received, that we shall make no apology for adopting it.

We are indebted to North-America for most of our Asters, but the present species, which is omitted by *Miller*, and is rather a scarce plant in this country, though not of modern introduction, being figured by *Plukenet* and described by *Ray*, is a native of Africa, and, like a few others, requires in the winter the shelter of a greenhouse.

It is particularly distinguished by having very narrow leaves with short bristles on them, and by its blossoms drooping before they open.

It is a perennial, flowers in September and October, and may be propagated by slips or cuttings.

The plant from whence our drawing was made, came from Messrs. *Gordon* and *Thompson*'s Nursery, Mile-End.

[Pg 074]

[34]

Browallia elata. Tall Browallia.

Class and Order.

Didynamia Gymnospermia.

Generic Character.

Cal. 5-dentatus. *Cor.* limbus 5-fidus, æqualis, patens: umbilico clauso Antheris 2, majoribus. *Caps.* 1-locularis.

Specific Character and Synonyms.

BROWALLIA *elata* pedunculis unifloris multiflorisque. *Lin. Syst. Vegetab. p.* 572. *Sp. Pl.* 880. *Mill. Dict.*

Of this genus there are only two species, both natives of South-America, the *elata*, so called from its being a much taller plant than the *demissa*, is a very beautiful, and not uncommon stove or green-house plant; it is impossible, by any colours we have, to do justice to the brilliancy of its flowers.

Being an annual, it requires to be raised yearly from seed, which must be sown on a hot-bed in the spring, and the plants brought forward on another, otherwise they will not perfect their seeds in this country. Some of these may be transplanted into the borders of the flower-garden which are warmly situated, where, if the season prove favourable, they will flower and ripen their seeds; but, for

security's sake, it will be prudent to keep a few plants in the stove or green-house.

As these plants have not been distinguished by any particular English name, Miller very properly uses its Latin one; a practice which should as much as possible be adhered to, where a genus is named in honour of a Botanist of eminence.

[Pg 075]

[Pg 076]

[35]

Crepis barbata. Bearded Crepis, or Purple-eyed Succory-Hawkweed.

Class and Order.

Syngenesia Polygamia Æqualis.

Generic Character.

Recept. nudum. *Cal.* calyculatus squamis deciduis. *Pappus* plumosus, stipitatus.

Specific Character and Synonyms.

CREPIS *barbata* involucris calyce longioribus: squamis setaceis sparsis. *Lin. Syst. Vegetab.* p. 719.

HIERACIUM proliferum falcatum. *Bauh. Pin.* 128.

HIERACIUM calyce barbato. *Col. ecphr.* 2. *p.* 28. *t.* 27. *f.* 1.

HIERACIUM boeticum medio nigro. *Herm. Parad. Bat.* 185. *t.* 185.

Grows spontaneously in the south of France, about Montpelier; also, in Spain, Italy, Sicily, and elsewhere in the south of Europe: is one of the most common annuals cultivated in our gardens. It begins flowering in July, and continues to blossom till the frost sets in.

No other care is necessary in the cultivation of this species than sowing the seeds in the spring, in little patches, on the borders where they are to remain, thinning them if they prove too numerous.

Miller calls this species *bœtica*, and improperly describes the centre of the flower as black, as also does Herman: in all the specimens we have seen, it has evidently been of a deep purple colour, or, as Linnæus expresses it, *atropurpurascens*.

[Pg 077]
[Pg 078]
[Pg 079]

[36]

Lilium bulbiferum. Orange Lily.

Class and Order.

Hexandria Monogynia.

Generic Character.

Cor. 6-petala, campanulata: *linea* longitudinali nectarifera. *Caps.* valvulis pilo cancellato connexis.

Specific Character and Synonyms.

LILIUM *bulbiferum* foliis sparsis, corollis campanulatis erectis: intus scabris. *Lin. Syst. Vegetab. p.* 324. *Jacq. Fl. Austr. t.* 226.

LILIUM purpureo-croceum majus. *Bauh. Pin.* 76.

LILIUM aureum, the gold red Lily. *Park. Parad. p.* 37.

"The common orange or red Lily is as well known in the English gardens as the white Lily, and has been as long cultivated here. This grows naturally in Austria and some parts of Italy. It multiplies very fast by offsets from the roots, and is now so common as almost to be rejected; however, in large gardens these should not be wanting, for they make a good appearance when in flower if they are properly disposed; of this sort there are the following varieties:

The orange Lily with double flowers,

The orange Lily with variegated leaves,

The smaller orange Lily.

These varieties have been obtained by culture, and are preserved in the gardens of florists. They all flower in June and July, and their stalks decay in September, when the roots may be transplanted and their offsets taken off, which should be done once in two or three years, otherwise their branches will be too large, and the flower-stalks weak. This doth not put out new roots till towards spring, so that the roots may be transplanted any time after the stalks decay till November. It will thrive in any soil or situation, but will be strongest in a soft gentle loam, not too moist." *Mill. Dict.*

Bears the smoke of London better than many plants.

Varies with and without bulbs on the stalks.

[Pg 080]
[Pg 081]

INDEX.

In which the Latin Names of the Plants contained in the *First Volume*, are alphabetically arranged.

Pl.

24 Agrostemma Coronaria.

10 Anemone *Hepatica*.

33 Aster tenellus.

34 Browallia elata.

17 Cactus flagelliformis.

27 Convolvulus tricolor.

13 Coronilla glauca.

35 Crepis barbata.

4 Cyclamen *Coum*.

7 Cynoglossum *Omphalodes*.

25 Dianthus chinensis.

12 Dodecatheon *Meadia*.

11 Erica herbacea.

5 Erythronium *Dens Canis*.

18 Geranium Reichardi.

20 Geranium peltatum.

3 Helleborus hyemalis.

8 Helleborus niger.

19 Hemerocallis flava.

31 Jasminum officinale.

1 Iris persica.

9 Iris pumila.

16 Iris variegata.

21 Iris versicolor.

30 Lilium chalcedonicum.

36 Lilium bulbiferum.

32 Mesembryanthemum dolabriforme.

6 Narcissus minor.

15 Narcissus *Jonquilla*.

22 Nigella damascena.

28 Passiflora cœrulea.

14 Primula villosa.

29 Reseda odorata.

2 Rudbeckia purpurea.

26 Stapelia variegata.

23 Tropæolum majus.

tredition®

tredition was established in 2006 by Sandra Latusseck and Soenke Schulz. Based in Hamburg, Germany, tredition offers publishing solutions to authors and publishing houses, combined with worldwide distribution of printed and digital book content. tredition is uniquely positioned to enable authors and publishing houses to create books on their own terms and without conventional manufacturing risks.

For more information please visit: www.tredition.com

TREDITION CLASSICS

This book is part of the TREDITION CLASSICS series. The creators of this series are united by passion for literature and driven by the intention of making all public domain books available in printed format again - worldwide. Most TREDITION CLASSICS titles have been out of print and off the bookstore shelves for decades. At tredition we believe that a great book never goes out of style and that its value is eternal. Several mostly non-profit literature projects provide content to tredition. To support their good work, tredition donates a portion of the proceeds from each sold copy. As a reader of a TREDITION CLASSICS book, you support our mission to save many of the amazing works of world literature from oblivion. See all available books at www.tredition.com.

 Project Gutenberg

The content for this book has been graciously provided by Project Gutenberg. Project Gutenberg is a non-profit organization founded by Michael Hart in 1971 at the University of Illinois. The mission of Project Gutenberg is simple: To encourage the creation and distribution of eBooks. Project Gutenberg is the first and largest collection of public domain eBooks.